PUT IT TOGETHER

by Patty Whitehouse

CONTENTS

Lots of Parts	2
Putting Materials Together	4
Soft and Hard	6
Springs	8
Moving Parts	10
Many Materials	12
Building Houses	14
Glossary/Index	16

rourkeeducationalmedia.com

Lots of Parts

Look around you. Do you see windows? Look closely at all of the parts of the windows. The parts are made from different materials. How many materials can you name?

This door is made from wood, metal, and glass.

Different materials are used to make a back pack. The back pack needs to be strong but light.

Imagine if the windows were made of just one of those materials. If the windows were made of just wood or metal, you couldn't see through them. If the windows were made only from glass, they would be hard to open or close. The metal, wood, and glass all work together to make a useful window.

Putting Materials Together

A material's properties describe how it looks, feels, acts, or reacts. Each material has its own special properties. We put materials together to use all their special properties.

We know that the properties of metal and rubber are just right for bicycle wheels.

We make umbrellas from cloth, metal, and plastic. The cloth is soft and waterproof. The metal is flexible and strong. The plastic is light and smooth. Together these materials make a great umbrella.

Soft and Hard

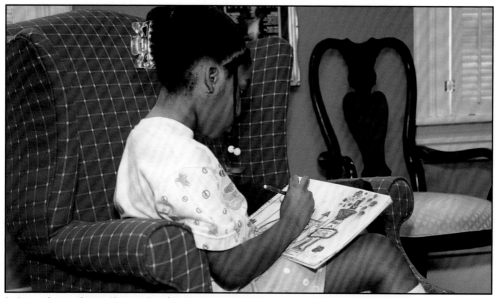

Wood makes the chair sturdy. Cotton stuffing makes the chair soft.

Some materials are soft. Some materials are hard. We use them together when we want to make something soft but sturdy. We make furniture from soft and hard materials.

Every day we use things made with hard and soft materials. A hard wooden pencil holds a soft eraser. A hard toothbrush holds soft bristles.

Springs

Coil springs are long pieces of wire wound into a tube shape. The most important property of a spring is that it is bouncy. It springs back into its original shape, even after it is pressed or stretched.

A trampoline uses coil springs to work. When you jump on a trampoline, the springs stretch, and the trampoline goes down. Then the springs pull back into their unstretched shape, and the trampoline pushes you up into the air.

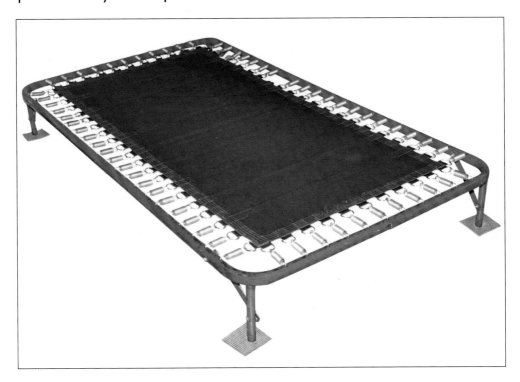

Moving Parts

Many things are put together so that some of their parts move. A bicycle has many moving parts. You move the pedals. They move the gears and chain that move the wheels. The bicycle can go because of its moving parts.

1. A rider pushes on the pedals
2. The pedal moves the gear
3. The gear moves the chain
4. The chain turns the wheels.

Many Materials

We use many things that help us live or amuse us. For example, we use washing machines and televisions. They are made of many different materials. Each material is used for its special properties. That is why the machines work.

Even a simple light bulb has many parts and materials. It has a glass bulb that keeps air from getting in and lets light out. It has a thin metal wire inside the bulb that gets so hot from electricity that it glows white. A bulb has a metal base that can be screwed into a lamp socket. The metal base conducts electricity to the glow wire.

Building Houses

We use many different materials to build a house.

Material	Property	Use
brick	Strong, waterproof	Covers the outside
wood	Strong	Makes the frame for the walls inside
glass	Transparent	Windows
copper	Conducts electricity	Wires for electricity
plastic foam	Does not conduct heat	Wall and ceiling insulation
Plastic tubes	Waterproof, will not rust	Pipes for plumbing

Properties and uses

Each material has a property that makes it useful for the building. When we put these materials all together, we have a safe and comfortable place to live!

GLOSSARY

glows	gives off light
materials	stuff we use to make or build things
socket	a hollow into which something fits
sturdy	strongly built
wound	wrapped around and around

INDEX

bicycle   4, 10, 11
light bulb   13
machines   12
properties   4, 8, 12, 15

umbrella   5
window   2, 3
wire   8, 13
wood   2, 3, 6, 7